WHAT DOES IT LOOK LIKE?

IT HAS SCALES!

BY KATHLEEN CONNORS

Gareth Stevens PUBLISHING

first concepts

Some animals have scales. Scales keep an animal's skin safe.

Reptiles have scales.
Snakes are reptiles.
They have scales.

Crocodiles have scales.

Turtles have scales.

Iguanas have scales.

Some fish have scales.
Goldfish have scales.

Sunfish have scales.

Some bugs have scales.
Butterflies have scales.

Moths have scales.

Pangolins are
mammals.
They have scales.

Which animal has scales?

Please visit our website, www.garethstevens.com. For a free color catalog of all our high-quality books, call toll free 1-800-542-2595 or fax 1-877-542-2596.

Cataloging-in-Publication Data
Names: Connors, Kathleen.
Title: It has scales! / Kathleen Connors.
Description: Buffalo, NY : Gareth Stevens Publishing, 2026. | Series: What does it look like?
Identifiers: ISBN 9781482470932 (pbk.) | ISBN 9781482470949 (library bound) | ISBN 9781482470956 (ebook)
Subjects: LCSH: Scales (Reptiles)–Juvenile literature. | Scales (Fishes)–Juvenile literature. | Body covering (Anatomy)–Juvenile literature.
Classification: LCC QL942.C747 2026 | DDC 591.47′7–dc23

Published in 2026 by
Gareth Stevens Publishing
2544 Clinton Street
Buffalo, NY 14224

Designer: Andrea Davison-Bartolotta
Editor: Kristen Nelson

Photo credits: Cover, p.1 (background) olgers/Shutterstock.com; cover, pp. 1 (iguana), 11, 23 (frog) Kurit afshen/shutterstock.com; p. 3 (bottom) Andreas Muth-Hegener/Shutterstock.com; p. 3 (top left) Charles T. Peden/Shutterstock.com; p. 3 (top right) Marc Freebrey/Shutterstock.com; p. 5 Agus_Gatam/Shutterstock.com; p. 7 Mary Angela Heys/Shutterstock.com; p. 9 Ihsan Sercan/Shutterstock.com; p. 13 ArtEvent ET/Shutterstock.com; p. 15 Griffin Gillespie/Shutterstock.com; p. 17 (main) James W. Thompson/Shutterstock.com; p. 17 (inset) Cornel Constantin/Shutterstock.com; p. 19 (main) Harvy Matters/Shutterstock.com; p. 19 (inset) bt_photo/Shutterstock.com; p. 21 Vickey Chauhan/Shutterstock.com; p. 23 (cat) Le Dang Nhut Minh/Shutterstock.com; p. 23 (fish) Tanveer Anjum Towsif/Shutterstock.com; p. 23 (pig) photomaster/Shutterstock.com.

Printed in China

CPSIA compliance information: Batch #QSGS26: For further information contact Gareth Stevens at 1-800-542-2595.